DiscoverRoo
An Imprint of Pop!
popbooksonline.com

How Animals Evolved

HOW REPTILES EVOLVED

by Grace Hansen

WELCOME TO DiscoverRoo!

This book is filled with videos, puzzles, games, and more! Scan the QR codes* while you read, or visit the website below to make this book pop.

popbooksonline.com/ev-reptiles

abdobooks.com

Published by Pop!, a division of ABDO, PO Box 398166, Minneapolis, Minnesota 55439.

Printed in the United States of America, North Mankato, Minnesota.

102023
012024

THIS BOOK CONTAINS RECYCLED MATERIALS

Cover Photo: Shutterstock
Interior Photos: Shutterstock; Getty Images; Science Source
Editor: Elizabeth Andrews
Series Designer: Laura Graphenteen

Library of Congress Control Number: 2023939060

Publisher's Cataloging-in-Publication Data

Names: Hansen, Grace, author.
Title: How reptiles evolved / by Grace Hansen
Description: Minneapolis, Minnesota : Pop!, 2024 | Series: How animals evolved | Includes online resources and index
Identifiers: ISBN 9781098245467 (lib. bdg.) | ISBN 9781098246020 (ebook)
Subjects: LCSH: Evolution--Juvenile literature. | Reptiles--Juvenile literature. | Reptiles--Evolution--Juvenile literature. | Biology--Juvenile literature.
Classification: DDC 591.38--dc23

*Scanning QR codes requires a web-enabled smart device with a QR code reader app and a camera.

TABLE OF CONTENTS

CHAPTER 1

WHAT IS EVOLUTION?

Evolution is the scientific **theory** that all life on earth has a common ancestor that lived billions of years ago. Living things **developed** from earlier versions of that ancestor. The changes that happened took millions of years.

WATCH A VIDEO HERE!

Life forms on Earth have evolved for millions of years.

Iguanas have adaptations that allow them to fight off predators. They are quick, have strong jaws, and sharp tails and teeth.

Evolution happens because of natural selection. This process occurs when one individual of a **species** has different **traits** from others. Sometimes the differences make surviving easier for the individual. As they survive, they pass on their traits to their young. Over time, more and more individuals in the species are born with the new and helpful trait until it is normal.

DID YOU KNOW?

Almost all scientists believe in evolution because it is testable.

Evolution is how fish eventually moved onto land. It is how birds began to fly and mammals started giving birth to live young. It is how humans began walking on two feet.

Only birds and bats have evolved to fly.

Iguanas are reptiles that have lived for more than 4.5 million years.

CHAPTER 2

WHAT ARE REPTILES?

Reptiles come in a wide array of sizes, shapes, and **traits**. They live in a variety of **ecosystems** around the world.

LEARN MORE HERE!

Gharials (top) *have unique snouts compared to other crocodilians.*

There are more than 150 chameleon species in the world. They can be found in Europe, Africa, and Asia.

The skin of crocodilians
is tough and protective.

There are four orders of reptiles. The order Crocodilia includes crocodiles, alligators, caimans, and gharials. Lizards and snakes make up the order Squamata. Tortoises, turtles, sea turtles, and other animals belong to the order Testudines. Finally, the lizard-like tuatara is alone in its order.

> **LONG LIVE REPTILES!**
>
> Two kinds of reptiles have very long lifespans. The tuatara can live up to 100 years. Giant tortoises can live well over 100 years!

As different as these reptiles are, they share common traits. Their skin is dry but may look shiny. Reptiles are covered in scales, bony plates, or both. Scales and plates make a reptile's skin waterproof.

Turtles can often be seen bathing in the sun.

Most reptiles cannot **regulate** their body temperature internally. They rely on the sun's rays to warm them and shade or water to cool them.

Reptile young hatch from eggs. Most reptiles lay soft-shelled eggs in nests that are then buried. This keeps the eggs safe until the young are ready to hatch.

Milk snakes are often confused with dangerous coral snakes and copperheads due to their coloring.

CHAPTER 3

EARLY REPTILES

The earliest known reptile, *Hylonomus lyelli*, lived approximately 350 million years ago (MYA) in what is now Canada. It looked like a lizard. It was eight inches (20cm) long. Slowly, over millions of years, *H. lyelli* and its **descendants** evolved into the reptiles that populate Earth today.

EXPLORE LINKS HERE!

Fossils of H. lyelli *remains have been discovered in fossilized trees in Nova Scotia, Canada.*

Apatosaurus *was a sauropod dinosaur that lived in North America about 151 million years ago.*

Dinosaurs were giant reptiles that first appeared 245 to 233 MYA. The exact timing of the evolution of dinosaurs is still being researched. About 65 MYA, all of the **non-avian** dinosaurs died out. The avian dinosaurs continued to evolve into birds. The mass extinction meant that other plants and animals had the opportunity to thrive. This led to the evolution of many new **species**.

DID YOU KNOW?

A group of dinosaurs called sauropods had very long necks. They could reach food high in trees.

CHAPTER 4

WORLD OF REPTILES

Tuataras are approximately 20 inches (50cm) long. They have evolved some unique features. The animals have a third eye on top of their heads. Scientists believe they use this eye to sense light.

COMPLETE AN ACTIVITY HERE!

Saltwater crocodiles are the largest and heaviest of today's reptiles.

Tuataras also have a unique set of teeth that help them crunch up insects and other **prey**.

Tuataras can be found in New Zealand.

SPECIALIZED CHAMELEONS

No other animals have shells with the same structure as a turtle's shell. Scientists think that shells may have evolved from ribs that had **broadened** out. Several fossils of ancient wide-ribbed animals have been found. These animals would have walked very slowly. The shell would have been a needed armor.

The Indian star tortoise is easily recognized by its beautiful shell.

Crocodilians' ability to be mostly submerged in water helps them hunt.

Today, there are 23 species of crocodilians swimming in fresh and salt waters around the world. They have powerful jaws and cone-shaped teeth. They have four short legs ending in feet with webbed toes. Their eyes and nose can stay above water while the rest of their heads and bodies are submerged.

Cone-shaped teeth are good for holding onto slippery fish!

Snakes with bright colors, such as the eyelash viper of Central and South America, are often venomous.

More than 3,700 species of snakes have been identified. However, very little is known about their evolution. Despite their lack of limbs, snakes move easily because of their **adapted** scales. A single row of ventral scutes runs the length of every snake's belly.

RATTLESNAKES

There are around 50 species of rattlesnakes. They have adapted many special ways to survive. Their coloration helps them hide. Their rattle makes a buzzing sound that warns **predators** to stay away. Their last line of defense is their venomous bite.

A rattlesnake takes a defensive posture.

Tokay geckos are brightly colored, but can change the color of their skin to blend in with their environment.

The around 1,700 species of geckos have adapted to live in nearly every kind of environment. They are found on every continent except Antarctica. Some have adapted to store fat in their tails. They use these fat stores when there is little food. Many have special toe pads that make them great climbers.

As research and discoveries add up, more information on reptiles will become known. Scientists work tirelessly to create an accurate picture of the evolution of all reptiles.

Geckos can walk up walls and across ceilings.

DID YOU KNOW?

Geckos have evolved a way to escape from predators. If a predator grabs a gecko's tail, the tail breaks off. It will later regrow!

MAKING CONNECTIONS

TEXT-TO-SELF

What part of the evolution of reptiles surprised you the most? Please explain your answer.

TEXT-TO-TEXT

Have you read any other books about animal evolution? What did the evolution of those animals have in common with reptile evolution?

TEXT-TO-WORLD

There are thousands of species of reptiles in the world. With the help of an adult, look up information about your favorite reptile. Write a short paragraph explaining why it is your favorite.

GLOSSARY

adapted — suited by nature or design to a particular use, purpose, or situation.

broaden — to make or become wide or wider.

descendant — an animal that comes from a particular ancestor or group of ancestors.

develop — to grow over time.

ecosystem — a community of organisms and their surroundings.

non-avian — not of or relating to birds.

predator — an animal that lives by hunting and eating other animals.

prey — an animal that is hunted by another animal for food.

regulate — to make accurate by adjusting.

species — a group of living things that are very much alike.

theory — an explanation for how things work and why things happen.

trait — a quality or feature of something.

INDEX

DiscoverRoo!
ONLINE RESOURCES

This book is filled with videos, puzzles, games, and more! Scan the QR codes* while you read, or visit the website below to make this book pop.

popbooksonline.com/ev-reptiles

*Scanning QR codes requires a web-enabled smart device with a QR code reader app and a camera.